BROTHER ALFRED BROUSSEAU

PROBLEM-SOLVING AND MATHEMATICS COMPETITION

INTRODUCTORY DIVISION

Lyle Fisher • Bill Kennedy

DALE SEYMOUR PUBLICATIONS

Cover Design by Margaret Sanfilippo

Copyright ©1984 by Dale Seymour Publications. All rights reserved. Printed in the United States of America. Published simultaneously in Canada.

Limited Reproduction Permission: The publisher grants permission to reproduce up to 100 copies of the worksheet pages in this book for noncommercial classroom or individual use. Any further duplication is prohibited.

Order number DS01447

ISBN 0-86651-228-4

DALE
SEYMOUR
PUBLICATIONS
P.O. BOX 10888
PALO ALTO, CA 94303

cdefghij-MA-8987

PREFACE

We believe that problem solving should be at the core of all mathematics curricula. We also believe that problem solving can and should be stimulating, challenging, and fun. Whether or not it is depends on the teacher's interest and skill in presenting problems and establishing an environment conducive to creative thinking. The essential condition for attaining this is that teachers must be problem solvers, and we suggest that the users of this book who first solve the problems will attain the maximum benefit.

Only by solving the problems will teachers fully appreciate what they are requiring of their students. Working through the problems will alert teachers to the skills required and may reveal the pitfalls students will encounter. Furthermore, there is an indefinable spirit of enthusiasm generated by teachers who are, themselves, problem solvers.

Problem solving is an art; it requires a certain feel, or touch. Though we have elsewhere listed strategies that our students have found useful, we wish to caution teachers not to classify problems by type, other than in a broad and fluid way. We must provide opportunity for students to create their own solutions. (Indeed, the users of this book may discover simpler, more elegant solutions than those we have presented.)

We are becoming an information society and the very accumulation of that information generates new information. The problem solvers of the future will be those who can examine information and look for patterns that suggest solutions. The ability to generalize from specific data may be a better preparation for a student than the repetition of arithmetic skills. It is important to make mathematics vital and exciting, and problem solving is the approach.

Though mathematics may be a paper and pencil sport, when approaches to solutions demand long and tedious calculations, we encourage the use of calculators and computers.

The solutions and suggestions for the teacher are the result of our work with students in grades 7-10 at Redwood High School, Larkspur, California; Kent Middle School, Kentfield, California; and the after-school enrichment classes sponsored by the Marin Education Task Force. It is these students that we wish to thank for their interest, creativity, perseverance, and curiosity. We also wish to thank our colleague, Bill Medigovich, for his encouragement and support.

Finally, this project would never have been completed without the efforts of Connee Fisher, who typed the entire manuscript and provided invaluable help with the editing.

L.F.
B.K.
San Rafael, California

INTRODUCTION

This book is a collection of 41 problems selected from the Brother
Alfred Brousseau Mathematics and Problem-Solving Competition. This
program provides high school and junior high school students with an
opportunity to solve problems of some difficulty not just once a
year, at the time of a mathematics contest, but throughout the year
on a continuing basis. The aims of the competition are:

1. To arouse mathematical interest and enthusiasm, especially
    in the key areas of problem solving and originality.
2. To bring to light mathematical talents that otherwise might
    remain hidden.
3. To help students make connections among widely separated
    areas of mathematics.
4. To direct abler students to advanced techniques and useful
    habits of generalization and abstraction.
5. To upgrade student performance in homework, examinations,
    and contests.
6. To supply teachers with diversified resource material.

The problems we have selected require various strategies. Some that
the students will find helpful are:

To look for a pattern
To guess and check
To draw a picture or make a model
To solve a simpler, related problem
To work backwards
To make an organized list, a table, or a tree diagram

Problem solving should be a part of every student's daily work. One
way is to provide from ten to fifteen minutes of each class period
for discussion and work on problems. Very few problems are completed
in a single period, but this gives students time to digest the problems, think about them, leave them alone, and come back with new,
fresh ideas.

Here are some suggestions that are essential for anyone who is
planning to teach problem solving:

Work the problems. Before assigning a problem, work it
yourself.
Define the problems. Carefully discuss the intent of a new
problem.
Require records of work. Students should record all
attempts--failures, as well as successes and write out
explanations. (We require that this work be kept in a
spiral notebook.)
Allow students to devise their own plans. Encourage group
effort.
Answer questions with questions. Don't give free information.
Take time. Give students the chance to play around with a
problem.

Mathematics Strands

The 41 problems that follow include practice in all of the strands
of the California State Mathematics Framework. Some of the problems
fall under more than one category, of course. Many of the problems
could be included as practice in the technology strand, but a
calculator or computer is especially useful in problems 1,2,4,5,6,
12,21,32, and 39.

NUMBER THEORY   PROBABILITY AND STATISTICS

1,5,9,10,12,13,20,   3,7,11,15,19,23,28
21,24,25,26,31,32,
36,38

GEOMETRY   MEASUREMENT   FUNCTIONS/LOGICAL REASONING

2,6,10,14,16,   2,6,13,16,17,   4,8,13,17,20,29,37
18,22,27,29,   22,27,30,33,
30,34,35,37,   34,35,39,41
39,40

Organization

The 41 problems in this book are presented on worksheet forms that
you can copy and hand out to your students. Following the work-
sheets, you will find detailed discussions for each problem
including suggested uses for the problem, complete solutions, and
teaching ideas. These discussions begin on page 42.

HISTORY OF THE
BROTHER ALFRED BROUSSEAU
PROBLEM-SOLVING AND MATHEMATICS COMPETITION

In 1959, long before others saw the importance of teaching problem solving, Brother Alfred Brousseau of St. Mary's College, Moraga, California, recognized the need. He began a problem-solving and mathematics competition for junior and senior high students. This year we will celebrate the twenty-fifth anniversary of that event.

The competition is unique in that the core of the program extends over the entire school year. Students are given a set of eight to ten problems four times during the year and are allowed ample time for experimentation and research. They must not only find a reasonable answer, they must also develop a logically defensible process for solving the problem. Answers are not accepted unless accompanied by a fully developed solution.

Based on their teacher's evaluation of their performance during the year, students may participate in the Finals Competition held the first Saturday in May at the University of San Francisco and at the University of California, Los Angeles. In 1983 more than 4500 high school and middle school students participated in the year-long problem-solving program and more than 400 attended the Finals..quite a contrast with the 20 students who competed in 1959.

When Brother Alfred Brousseau and his colleague at St. Mary's, Brother Brendan Kneale, initiated the program, its purpose was to stimulate math achievement in the nine high schools of the Christian Brothers of California. A statement as to the nature of the competition was made that year. "The ratings of problems will take into account the following factors: (1) correctness of the solution; (2) its brevity and elegance; (3) neatness of presentation."

After two successful years, the program was expanded. The mathematics department at The College of the Holy Names, led by Sisters Madeline Rose Ashton and Rose Eleanor Ehert, joined the effort. That third year, seventeen schools participated.

In 1962-63 the competition grew rapidly. The program received financial assistance from the California Mathematics Council, Northern Section, and from the National Science Foundation, in the form of a one-year grant. In that year, thirty-nine schools and nearly 1000 students participated.

The school year 1963-64 saw the development of the basic format that is still used today. Two separate problem-solving programs were announced: one for grades 7-9, and the other for grades 10-12. This gave greater emphasis to the junior high level and allowed the program to expand its scope. Four problem sets were distributed to participating schools on specific dates and were returned to St. Mary's and Holy Names' staff members for grading. At the end of the year there was an hour-long examination in each division. This was the final year of sponsorship by Holy Names.

In the next few years the Competition expanded to nearly 100 schools. The announcement of the 1969-1970 Competition contained this significant point: "As an experiment, this year we are (gratefully) accepting the offer of one school (Redwood High of Marin County) to assist us by forming a committee to correct the student papers of their school and to send the scores to us for distribution in our report." The program was simply growing too large for the St. Mary's staff to handle all the grading (more than 2000 students were involved at this point). That problems were corrected elsewhere was not considered critical, as this simply provided entrance into the Finals Competition, and the next year all schools were asked to correct their students' papers.

At this point, in 1975, St. Mary's College announced that it was discontinuing the Competition. For sixteen years the program had been carried almost entirely by Brothers Alfred Brousseau and Brendan Kneale. Though they must have derived great satisfaction from knowing that the Competition had encouraged thousands of students and hundreds of teachers, it was time to pass on the responsibility of managing the program.

Many teachers expressed regret that this fine program was ending, and two of them, Lyle Fisher and William Medigovich, of Redwood High School, Larkspur, California, stepped forward to shoulder the load. They were encouraged by Brother Alfred to continue the Competition.

Initial costs of continuing the program were paid by Redwood's district, the Tamalpais Union High School District. Brother Alfred, as he had for the previous sixteen years, and as he continues to do to this day, contributed problems. Many teachers provided problems and ideas, and by November 1975, the program was again in full swing with 91 schools and nearly 2500 students involved. The program was re-named the Brother Alfred Brousseau Problem-Solving and Mathematics Competition.

The 1976 Finals Competition was held, for the first time, at the University of San Francisco, thanks to the efforts of Dr. Millianne Lehmann. Dr. Lehmann and the University have continued to act as sponsor.

The publicity for the Competition in California Mathematics Council periodicals sparked the interest of a few schools in southern California. Two of the schools raised money and attended the Finals Competition. As a consequence of this, the California Mathematics Council decided to seek greater involvement of the southern California schools. In 1980, William Medigovich, CMC-N Treasurer, and Lyle Fisher, CMC President, proposed that the state organization assume the sponsorship of the Competition. This was approved, and the Brother Alfred Brousseau Problem-Solving and Mathematics Competition became an official statewide activity.

By 1981, interest among southern California schools was
sufficient to warrant the establishment of a Finals Competition
there. Joan Gell, of Palos Verdes High School, assumed the role
of program coordinator, and on May 2, 1981, a Finals Competition
was held at the University of California, Los Angeles, concurrent
with the one at the University of San Franciso.

ABOUT BROTHER ALFRED BROUSSEAU

It was with great pride that we named the Competition for
Brother Alfred Brousseau, our distinguished mentor and colleague
in the field of mathematics education. A lifelong scholar,
Brother Alfred has earned great distinction for his achievements
as an educator, an administrator, and as a moving force in the
development of important and innovative material for the continuing
education of teachers.

During Brother Alfred's more than fifty years of secondary
and college teaching in Christian Brothers schools, he has written
extensively and has broken much new ground in the concept of
mathematics as a stimulating, enjoyable pursuit comparable to
music, art, and literature. Most important, he has shared this
vision generously with his colleagues and with students.

Brother Alfred's academic credentials include bachelor's and
master's degrees in mathematics and a Ph.D. in physics, all from
the University of California, Berkeley. His professional affilia-
tions include the California Mathematics Council, in which he has
served with distinction as State President, as well as Treasurer
and President of the Northern Section. As professor of mathematics
at St. Mary's College, Moraga, Brother Alfred has been in constant
demand as a speaker at mathematics conferences and at high school
programs. He conceived and created the St. Mary's Problem Solving
and Mathematics Competition.

Brother Alfred has richly earned the honor of having the
Competition bear his name, for it is his vision and creativeness
that have made the Competition a vital force in mathematics
education in California.

PROBLEM 1

a) Using the table below, list the three-digit multiples of eleven. Fill in the table by columns, from top to bottom.

b) What patterns do you observe in the table?

c) Which two of these multiples satisfy this condition: division by eleven results in a quotient that is equal to the sum of the squares of the digits? For example, look at 132. Division by eleven gives a quotient of 12. The sum of the digits is 1 + 3 + 2 = 6; therefore, 132 does not meet the condition.

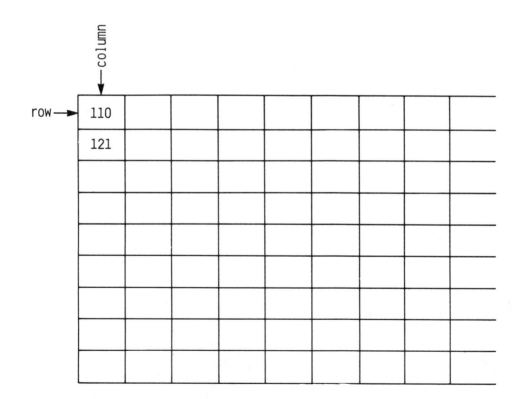

PROBLEM 2

The dimensions of a rectangular box are in the ratio of 2:3:5 and its volume is 82,320 cm^3. Find the dimensions of the box.

Copyright ©1984 by Dale Seymour Publications

PROBLEM 3

Given a regular set of dice, determine the probability of rolling each of the sums 2 - 12.

Fill in the tables.

2nd Die

+	1	2	3	4	5	6
1						
2						
3						
4						
5						
6						

1st Die

Sum of dice (n)	2	3	4	5	6	7	8	9	10	11	12
Probability of sum (P(n))	$\frac{1}{36}$	$\frac{2}{36}$									

Given two unmarked, cubic dice, and using the set of whole numbers, place numbers on each die such that only the following sums can be thrown: 5,7,9,11,13,15,17,19,21,23,25,27, and the probability of rolling each sum is the same. (There is more than one such set. Can you find others?)

Copyright ©1984 by Dale Seymour Publications

Introductory Division

PROBLEM 4

There is a function machine that operates on a given input number (and no other numbers) using addition, subtraction, multiplication, division, or any combination of these four operations. Complete the following table.

Input (x)	Output
1	1
2	9
3	29
4	67
5	129
6	221
7	
8	
9	
10	
.	
.	
.	
.	
x	

PROBLEM 5

Complete the following tables:

a)

n	3^n
1	$3^1 = 3$
2	$3^2 = 9$
3	
4	
5	
6	
7	
8	
9	
10	

b)

n	7^n
1	
2	
3	
4	
5	
6	
7	
8	
9	
10	

c)

n	4^n
1	
2	
3	
4	
5	
6	
7	
8	
9	
10	

d) What is the units digit for the following sum?

$$13^{841} + 17^{508} + 24^{617} =$$

Copyright ©1984 by Dale Seymour Publications

Introductory Division

PROBLEM 6

The length of a rectangular parallelepiped is four times the height, and the width is twice the height. If a space diagonal is 42 cm, find the volume (to the nearer cubic centimeter). One of the space diagonals in the figure below is $\overline{AG}$. $\overline{AF}$ is not a space diagonal.

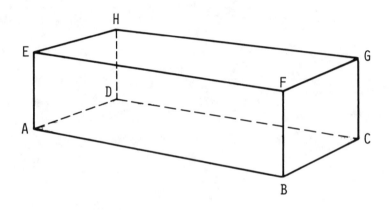

PROBLEM 7

VIBGYOR is a way of remembering the principal colors of the spectrum: violet, indigo, blue, green, yellow, orange, and red. Suppose there is a design with three square spaces in a row and that each of these spaces is to be filled with a different color from the above set. Furthermore, the colors must be in the same order as those given above. For example, VBR would be allowable, while GBR and GRB would not be allowable. In how many ways can the spaces be colored?

Copyright ©1984 by Dale Seymour Publications

PROBLEM 8

A_1 is a four by four square array of the natural numbers 1 to 16. A_2 represents a transformation of this array.

$$A_1 = \begin{array}{cccc} 1 & 2 & 3 & 4 \\ 5 & 6 & 7 & 8 \\ 9 & 10 & 11 & 12 \\ 13 & 14 & 15 & 16 \end{array} \qquad A_2 = \begin{array}{cccc} 7 & 1 & 5 & 11 \\ 6 & 2 & 3 & 8 \\ 14 & 10 & 16 & 13 \\ 9 & 15 & 4 & 12 \end{array}$$

Note that, in this transformation the quantity in position $1(A_1)$ moves to position $2(A_2)$; the quantity in position $2(A_1)$ moves to position $6(A_2)$; the quantity in position $3(A_1)$ moves to position $7(A_2)$; the quantity in position $4(A_1)$ moves to position $15(A_2)$; etc. Thus,

$$A_3 = \begin{array}{cccc} 3 & 7 & 6 & 16 \\ 2 & 1 & 5 & 8 \\ 15 & 10 & 12 & 9 \\ 14 & 4 & 11 & 13 \end{array}$$

How many transformations are required to arrive back at the original arrangement of A_1? Explain how you arrived at your answer.

Copyright ©1984 by Dale Seymour Publications

Introductory Division

PROBLEM 9

Four positive integers a, b, c, d, are such that a and b have a common factor greater than one; a and c have a common factor greater than one; a and d have a common factor greater than one; b and c have a common factor greater than one; b and d have a common factor greater than one; c and d have a common factor greater than one. Upon adding the four numbers a + b + c + d we have a sum that is prime. Find the set of the smallest integers for which this is true.

PROBLEM 10

a) Given a set of points in a plane, no three of which are
    collinear (that is, no three are in a line), how many line
    segments can be formed?

(1)   •   (2)   • •   (3)   • •   •   (4)   • •
                                     •   •   •

(5)   •   •   (6)   •   •
                         •   •
                           •

b) Complete the following table:

No. of non-collinear points	1	2	3	4	5	6	7	8	9	10
No. of segments formed										

c) State a formula that will produce the set of segments in
    Part b.

         S =

Copyright © 1984 by Dale Seymour Publications

PROBLEM 11

A straight in poker consists of five cards in sequence regardless of their suit. A straight flush consists of five cards in sequence and all of the same suit. If the ace can be used either as a high or low card, how many simple straights (not including straight flushes) are there? Explain.

Copyright ©1984 by Dale Seymour Publications

PROBLEM 12

Complete the following and you will see an interesting pattern emerge.

12345679 x 9 =

12345679 x 18 =

12345679 x 27 =

12345679 x 36 =

12345679 x 45 =

12345679 x 54 =

12345679 x 63 =

12345679 x 72 =

12345679 x 81 =

Look at the first multiplication, 12345679 x 9 = 111111111. From the product at the right determine two prime factors of 12345679.

PROBLEM 13

If it is 11:11 AM, what time will it be 143,999,999,993 minutes later?

Copyright ©1984 by Dale Seymour Publications Introductory Division

PROBLEM 14

a) In the five-sided prism shown below:

    (1) Identify the lower base.

    (2) Identify the upper base.

    (3) Identify a lateral edge.

    (4) Identify a lateral face.

    (5) What relationship holds for the upper base and lower
         base?

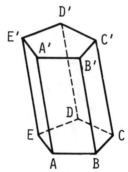

b) An example of a space diagonal in the prism pictured above
    would be $\overline{AD}'$. How many space diagonals are contained in a
    30-sided prism? (i.e., one whose upper and lower bases
    have 30 sides each.)

c) Can you determine a formula for the number of space
    diagonals in an n-sided prism?

Copyright ©1984 by Dale Seymour Publications

PROBLEM 15

a) Using the digits 1, 2, 3, 4, and 5, how many five-digit positive integers can be formed if no digits may be repeated?

b) Using the digits 1, 2, 3, 4, and 5, how many five-digit positive integers can be formed if only the digit 5 may be repeated any number of times?

Copyright ©1984 by Dale Seymour Publications

PROBLEM 16

A bathysphere with a diameter of ten feet descends to a depth of 1000 feet. If sea water weighs 64.3 lbs per cubic foot, and we consider all parts of the bathysphere as being under the same pressure, what is the total force in tons (to the nearer unit) on the bathysphere at that depth?

Copyright ©1984 by Dale Seymour Publications

PROBLEM 17

A mathematics explorer had to cross the great Coordinate Plain, a distance of 800 miles, with no gasoline supply stops. His late-model Cartesian truck averaged ten miles to a gallon of gasoline. The truck's gas tank held only ten gallons, but the truck could also carry 40 more gallons in eight five-gallon cans. This gave the truck a total capacity of 50 gallons and a range of 500 miles. At his starting point he had an unlimited supply of gasoline and five-gallon cans. With this in mind, the intrepid mathematician determined a means of crossing the Coordinate Plain and the minimum number of gallons of fuel necessary for the trip. How did he do it, how many gallons of gasoline did he use, and how many miles did he travel?

Copyright ©1984 by Dale Seymour Publications

PROBLEM 18

A regular triangle (equilateral) has angles of 60°. A regular quadrilateral (square) has angles of 90°.

a) What is the measure of each angle in a regular pentagon? A regular hexagon? A regular heptagon? A regular octagon?

b) Complete the following table.

sides (n)	sum of all angles	measure of each angle
3		
4		
5		
6		
7		
8		
9		
10		
11		
12		
.		
.		
.		
n		

c) What is the measure of each angle in a polygon of 36 sides?

d) If each angle of a regular polygon is 171°, how many sides does the polygon have?

Copyright ©1984 by Dale Seymour Publications

PROBLEM 19

In a series of five games to be played by two equally matched teams, the team that wins three games first is the champion. If Team A has won the first game, what is the probability that Team A will win the championship?

PROBLEM 20

Let $f(m) = N$ be a function described as follows: m is the sum of the prime factors (remember, 1 is not a prime) included in N and N is the largest number for which the sum of the primes is m for any given m. For example, for m = 10, then $N = 36 = 2 \cdot 2 \cdot 3 \cdot 3$. (Refer to the table below.) There is no larger value for which the sum of the prime factors is 10.

Find N for (a) m = 12.

(b) m = 14.

Sums of primes = 10	Product of those primes
2 + 2 + 2 + 2 + 2	$2^5 = 32$
2 + 2 + 3 + 3	$2^2 \cdot 3^2 = 36$
2 + 3 + 5	$5 \cdot 3 \cdot 2 = 30$
5 + 5	$5^2 = 25$

Copyright © 1984 by Dale Seymour Publications   Introductory Division

PROBLEM 21

a) The sequence of Fibonacci numbers begins 1,1,2,3,5,8,13,21.... Note that the sequence begins with 1,1 and that each successive term is the sum of the two preceding terms. Thus, the third term, which we'll designate $F(3)$, is formed by adding $F(1)$ and $F(2)$, or $1 + 1 = 2$; $F(4) = 1 + 2 = 3$; $F(5) = 2 + 3 = 5$; $F(6) = 3 + 5 = 8$; etc. Using the table below, list the first 20 terms of the Fibonacci sequence.

*n	1	2	3	4	5	6	7	8	9	10	11	12	13	14	15	16	17	18	19	20
F(n)	1	1	2	3	5	8	13	21												

*n is the number of the term in the sequence.

b) Note that it is possible to work backwards within the sequence. For example, the 6th term, $F(6)$, is equal to $F(8) - F(7)$, or, in general terms $F(n) = F(n+2) - F(n+1)$. Using the table below, fill in the missing terms in the sequence.

n	5	4	3	2	1	0	-1	-2	-3	-4	-5	-6	-7	-8	-9	-10	-11	-12	-13	-14	-15
F(n)	5	3	2	1	1	0	1	-1													

c) State a formula for F_{-n}.

Copyright © 1984 by Dale Seymour Publications

PROBLEM 22

A regular star is formed by extending the sides of a regular
n-gon to a first point of intersection with every other side, as
in the figure below:

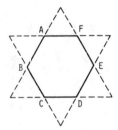

Thus, side $\overline{AF}$ is extended to intersect with the extensions of
sides $\overline{ED}$ and $\overline{BC}$. Side $\overline{CD}$ is extended to a point of intersection
with the extensions of sides $\overline{EF}$ and $\overline{AB}$. Continuing this process,
we form a regular six-pointed star.

a) What is the sum of the measures of the angles of the six-
    pointed star?

b) Complete the following table for all n-gons.

n vertices	measure of the star angles	sum of the star angle measures
5		
6		
7		
8		
9		
10		
.		
.		
.		
n		

Copyright ©1984 by Dale Seymour Publications

PROBLEM 23

Suppose you are playing a game in which three regular dice are used.

a) How many possible outcomes are there?

b) What are the most probable outcomes?

c) What is the probability of rolling a 9?

d) What is the probability that you will roll three different numbers?

e) What is the probability of rolling a sum less than 12?

Copyright ©1984 by Dale Seymour Publications

PROBLEM 24

For n is greater than 7, find three consecutive integers n, n+1, and n+2, such that these are divisible respectively by 5, 6, and 7. Let n be the smallest integer satisfying this requirement.

PROBLEM 25

For the following integers in Base 7 determine which are even and which are odd.

a)   1364253

b)   2130531

c)   1103245

d)   5314216

e)   3040312

How can you decide if a Base 7 integer is odd or even without changing it to Base 10?

A very difficult challenge is to determine a divisibility rule for 8 in Base 7. (Note that 8 is 11 in Base 7.) You might wish to compare patterns of this problem with those of PROBLEM 1.

Copyright ©1984 by Dale Seymour Publications   Introductory Division

PROBLEM 26

The table below is built by adding three elements at a time from the previous row. Row One is given:

Row One	1	1	1				
Row Two	1	2	3	2	1		
Row Three	1	3	6	7	6	3	1

We obtain Row Two by adding $0 + 0 + 1 = 1$; $0 + 1 + 1 = 2$; $1 + 1 + 1 = 3$; $1 + 1 + 0 = 2$; $1 + 0 + 0 = 1$. We obtain Row Three by adding $0 + 0 + 1 = 1$; $0 + 1 + 2 = 3$; $1 + 2 + 3 = 6$; $2 + 3 + 2 = 7$; $3 + 2 + 1 = 6$; $2 + 1 + 0 = 3$; $1 + 0 + 0 = 1$. Build the table up to seven rows.

a) How many terms are in each row? State a formula for the number of terms in Row Six. In Row N.

b) Find the sum of the elements in each row. What pattern do you observe?

   By algebra determine $(x^2 + x + 1)^1$, $(x^2 + x + 1)^2$,

   $(x^2 + x + 1)^3$. Looking at the table above, what will be the

   expansion of $(x^2 + x + 1)^6$?

PROBLEM 27

A ship drops anchor at slack water (i.e., when there is no current). Four hours later the tide has fallen 8 feet and the ship has moved a horizontal distance of 80 ft from its anchor. How many feet of anchor chain are in the water?

Copyright ©1984 by Dale Seymour Publications

PROBLEM 28

Connee is holding four cards: a ten of spades, a ten of hearts, a ten of diamonds, and a ten of clubs. Carol draws two of the cards. What is the probability that Carol will draw at least one red ten?

Copyright ©1984 by Dale Seymour Publications

PROBLEM 29

On a rectangular grid with coordinate axes starting at (0,0), carry out the following process. From the origin, go to the right one unit to (1,0); go to the left (up) two units to (1,2); go to the left three units to (-2,2). At each step you continue to change direction by 90 degrees to the left and move a distance one unit greater than in the previous step. What is the final position after 100 moves?

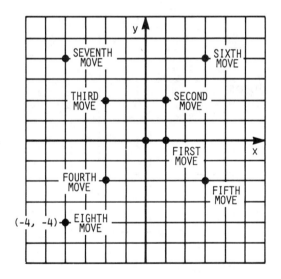

PROBLEM 30

A rectangle is constructed with adjacent sides of 10 centimeters and 8 centimeters. A second rectangle with adjacent sides of 7 centimeters and 5 centimeters is constructed and overlaps the first rectangle as shown in the figure below. What is the absolute value of the difference between the two non-overlapping regions of the rectangles?

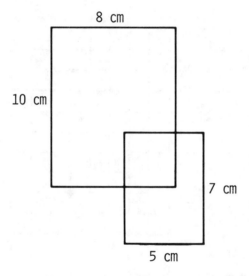

PROBLEM 31

A table is formed in the same way as the Pascal Triangle such that the first four rows are:

Row 1:   1   2

Row 2:   1   3   2

Row 3:   1   4   5   2

Row 4:   1   5   9   7   2

a) Build the table until there are ten lines.

b) Find the sum of the upward diagonals. (left to right)

c) Without building the table further, write the next five diagonal sums. Explain.

d) Find a formula for the sum of the quantities in the nth row.

Copyright © 1984 by Dale Seymour Publications   Introductory Division

PROBLEM 32

Given: $A = \sqrt{x + \sqrt{2x - 1}} + \sqrt{x - \sqrt{2x - 1}}$ and x is a real number.

a)   What is the minimum value that can be assigned to x?

b)   For what values of x is $A = \sqrt{2}$?

c)   If A = 2, what is the value of x?

Copyright ©1984 by Dale Seymour Publications

PROBLEM 33

There is a source of brine containing five pounds of salt per gallon. A tank of pure water has 1/3 of its contents drained off and the tank is then filled from the brine tank. This process is completed seven times. How much salt per gallon is in the liquid in the tank after these processes? (Express as the ratio of two integers $\frac{a}{b}$, in lowest terms.)

PROBLEM 34

A soup company decides to change the size of the cylindrical can in which its product is packaged. The marketing division decides that the new can should be one inch taller than the old can but should have the same volume. If the old cans had a height of 4 inches and a diameter of 2.6 inches, what should be the diameter of the new can? (round to hundredths)

Copyright © 1984 by Dale Seymour Publications   Introductory Division

PROBLEM 35

If the area of each circle in the figure below is $\frac{\pi}{4}$ and the distance between the centers of the circles P and Q is $\sqrt{3}$, what is the area of rectangle ABCD not covered by the circles? The circles are congruent, and are tangent to each other. The circles in the bottom row are tangent to $\overline{AD}$, $\overline{AB}$, and $\overline{BC}$, and the top circle is tangent to $\overline{CD}$. Also, $\overline{PQ}$ is parallel to $\overline{BC}$. Express your answer in terms of π and in simple radical form.

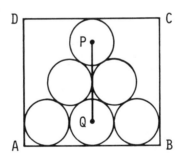

Copyright ©1984 by Dale Seymour Publications

PROBLEM 36

a) When a Base 10 integer is reversed and the difference found between this integer and the original, the difference is always divisible by what number?

b) For this difference, the sum of the digits is always divisible by what integer?

c) For the following Base 7 integers, reverse the digits and find the difference.

    436021   653012   540613

d) This difference is always divisible by what number?

e) For this difference, the sum of the digits is always divisible by what integer?

Copyright © 1984 by Dale Seymour Publications

PROBLEM 37

On a rectangular grid with coordinate axes, let a unit step be the diagonal of a unit square. Starting from the origin, go one step to (1,1). Then turn 90 degrees counterclockwise (to the left) and go two steps to (-1,3). Turn 90 degrees counterclockwise (to the left) and go three steps to (-4,0). At each step you continue to change by 90 degrees counterclockwise and increase the length of the vector by one at each step. What is the final position after 100 moves?

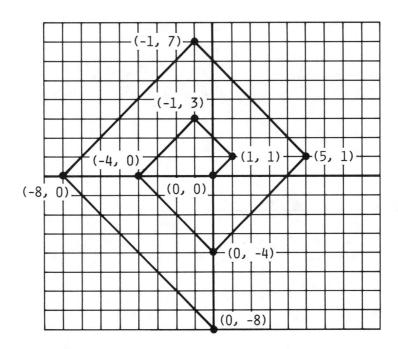

PROBLEM 38

In Base 10, 12345679 · 9 = 111111111 and 12345679 · 18 =
222222222.   Write similar relations for Bases 4, 5, 6, 7, 8,
9, 11, and 12.   That is, find the factors that will give
products of 111, 1111, 11111, etc.   In Base 11, use A = 10.
In Base 12, use B = 11.   Show work for the second table.

Base	Relation
4	
5	
6	
7	
8	
9	
10	12345679 · 9 = 111111111
11	
12	

Base	Relation
4	
5	
6	
7	
8	
9	
10	12345679 · 18 = 222222222
11	
12	

Copyright ©1984 by Dale Seymour Publications
Introductory Division

PROBLEM 39

How many circular pipes with an inside diameter of one inch would be needed to carry the same amount of water as a pipe with an inside diameter of one foot?

Copyright ©1984 by Dale Seymour Publications

PROBLEM 40

A certain dart board is constructed as shown below. If the region within the inner circle counts 100 points, the band between the first and second circles is 50 points, and a dart in the outside band is 25 points, what are the radii of the two outer circles if the inner circle has a radius of 3 inches and the areas of the other two regions are inversely proportional to their point value? (Give your answer correct to two decimal places.)

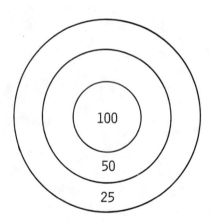

PROBLEM 41

Suppose I have $12,000 to invest. I place part of it in Bank A at 11% per year and the remaining part in Bank B at 8.5% per year. If I earn $1,257.50, in one year, how much did I deposit in each bank?

PROBLEM 1

ANSWER:

The two multiples that satisfy the condition are 803 and 550.
There are many observable patterns (see Solution, below).

USES OF THE PROBLEM:

This problem is good for illustrating the value of a table in
revealing patterns and organizing data. The second part of the
problem (finding the two multiples) is good practice in using logical analysis to eliminate choices, thus simplifying the problem.

110	209	308	407	506	605	704	803	902
121	220	319	418	517	616	715	814	913
132	231	330	429	528	627	726	825	924
143	242	341	440	539	638	737	836	935
154	253	352	451	550	649	748	847	946
165	264	363	462	561	660	759	858	957
176	275	374	473	572	671	770	869	968
187	286	385	484	583	682	781	880	979
198	297	396	495	594	693	792	891	990

SOLUTION:

Among the many patterns in the table, we see that:

   -looking across the rows, the units digit decreases by one
    until it reaches zero, then begins the cycle again at nine.

   -looking down the columns, the units digit increases by one
    until it reaches zero, then begins the cycle again at one.

   -there are similar patterns in the hundreds and tens digits.

   -the diagonals from upper left to lower right have a pattern
    of a constant units digit, and tens and hundreds digits that
    increase by one.

The two multiples that have a quotient on division by eleven equal
to the sum of the squares of the digits are 803 and 550. Looking
at the extremes of the multiples, 110 and 990, we see that
$110 \div 11 = 10$ and $1^2 + 1^2 + 0^2 = 2$; $990 \div 11 = 90$ and $9^2 + 9^2 + 0^2 =$
162. Thus, the lowest possibility for the sum of the squares of
the digits is 2 and the greatest is 90. Thus, any multiple with
a pair of nines will produce a sum of the squares of the digits
that is too great.

42

Likewise, we can eliminate multiples that have these combinations: (9,8), (9,7), (9,6), (9,5), (9,4), (8,8), (8,7), (8,6). All of these pairs give a sum of the squares of the digits greater than 90, the highest quotient. In this way, we eliminate 34 of the 81 possibilities. Similarly, 110 and 121 can be eliminated because their sums are less than 10, the lowest possibility for a quotient.

We might now consider 902, the greatest remaining number in the table. $902 \div 11 = 82$ and $9^2 + 0^2 + 2^2 = 85$. This eliminates 902, but we are close. Trying 825, we get $825 \div 11 = 75$ and $8^2 + 2^2 + 5^2 = 93$. Since we are taking successively lower values from the table, this last trial tells us the sum of the digits squared must be less than 75. This eliminates any number containing the pairs (8,5), (8,4), (7,6).

Our next trial is 803, and $803 \div 11 = 73$, while $8^2 + 0^2 + 3^2 = 73$. We have found one of the two possibilities.

Proceeding in similar fashion, we can find the other multiple, 550.

TEACHING SUGGESTIONS:

Though some students will solve the problem by "brute force" (trying every multiple until they find the two that work), others will discover "short-cuts." Encourage them to share their methods.

You might mention that this problem is a variation of one that was on the 1961 International Mathematics Olympiad, a world-wide competition among the top eight eighteen-year-olds from each of the participating countries.

PROBLEM 2

ANSWER:

28 cm x 42 cm x 70 cm

USES OF THE PROBLEM:

This problem offers practice in factoring, determining the volume of a rectangular prism, and the concept of ratio.

SOLUTION:

Let 2x, 3x, and 5x be the dimensions of the box.

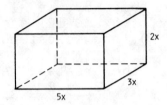

Then, $2x \cdot 3x \cdot 5x = 82320$

$$30x^3 = 82320$$

$$x^3 = 2744$$

Examining the prime factors of 2744 we find:

$$2744 = 2 \times 2 \times 2 \times 7 \times 7 \times 7$$

From this we can see that 2744 is a perfect cube:

$$2744 = 14 \times 14 \times 14$$

Thus, x = 14. The dimensions are:

$$2x = 28 \text{ cm}$$

$$3x = 42 \text{ cm}$$

$$5x = 70 \text{ cm}$$

TEACHING SUGGESTIONS:

Though the cube root of 2744 can be found using a calculator, encourage your students to develop a strategy for finding the root by factoring. Have them use this method to find the cube roots of 9261 (21), 74088 (42), and 50653 (37).

PROBLEM 3

ANSWER:

(a)

n	2	3	4	5	6	7	8	9	10	11	12
P(n)	$\frac{1}{36}$	$\frac{2}{36}$	$\frac{3}{36}$	$\frac{4}{36}$	$\frac{5}{36}$	$\frac{6}{36}$	$\frac{5}{36}$	$\frac{4}{36}$	$\frac{3}{36}$	$\frac{2}{36}$	$\frac{1}{36}$

(b)   One possibility for the unmarked dice is:

first die: 0, 2, 4, 6, 8, 10

second die: 5, 5, 5, 17, 17, 17

n	5	7	9	11	13	15	17	19	21	23	25	27
P(n)	$\frac{3}{36}$	$\frac{3}{36}$	$\frac{3}{36}$	$\frac{3}{36}$	$\frac{3}{36}$	$\frac{3}{36}$	$\frac{3}{36}$	$\frac{3}{36}$	$\frac{3}{36}$	$\frac{3}{36}$	$\frac{3}{36}$	$\frac{3}{36}$

USES OF THE PROBLEM:

This is a good introductory problem in probability because students chart all possible outcomes.

SOLUTION:

Merely filling in the tables shows the possible outcomes. Counting successful outcomes and dividing by possible outcomes gives the probability.

In the second part, one possible table would be:

+	5	5	5	17	17	17
0	5	5	5	17	17	17
2	7	7	7	19	19	19
4	9	9	9	21	21	21
6	11	11	11	23	23	23
8	13	13	13	25	25	25
10	15	15	15	27	27	27

A second possible set of dice could be marked as follows:

first die: 1, 3, 5, 7, 9, 11

second die: 4, 4, 4, 16, 16, 16

TEACHING SUGGESTIONS:

Remember, there are many possible sets for the second part. You might hold a contest to see who can find the greatest number of these sets.

You could extend the problem by changing the sums to be rolled, or by using dice that have other than six faces.

Another interesting extension would be to construct a seven-sided die that would provide a fair roll. (Hint: think of a die constructed so that it is similar to a pencil with seven sides and the ends coming to a point.)

For students without much background in probability, it is important to examine the table carefully and see how it accounts for all possible outcomes.

PROBLEM 4

ANSWER: See solution, below

USES OF THE PROBLEM:

Pattern analysis and inductive reasoning are the skills necessary to solve this problem.

SOLUTION:

Input (x)	Output
1	1
2	9
3	29
4	67
5	129
6	221
7	349
8	519
9	737
10	1009
$\vdots$	$\vdots$
x	$x \cdot x \cdot x + x - \dfrac{x}{x}$

It can be seen that the output is close to the cube. If we subtract the cube for one, we have zero, for two we have one, for three we have two, for four we have three, for five we have four, and for six we have five. We can see that we are adding one less than the input number x^3. As we can only operate using the input number, we must subtract one in the form of $\frac{x}{x}$. Thus, the operation must be $x \cdot x \cdot x + x - \frac{x}{x}$.

TEACHING SUGGESTIONS:

You might introduce this problem with simpler function machine problems(also known as the "What's My Rule?" game). For example,

input (x)	output		input (x)	output
3	8		2	-4
4	11		3	-2
5	14		4	0
.	.		5	2
.	.		.	.
.	.		.	.
x	3x-1		10	12
			.	.
			.	.
			x	2x-8

A good way to do this is to write the words "input" and "output" on the chalkboard or overhead projector. Ask students to give the input numbers and you write the output, using whatever rule you have in your head. Don't allow them to call out the rule; rather, when they think they have it, let them give the output for another student's input. Start with simple examples for which the general rule is obvious.

One extension of the problem is to determine the formula through finite differences.

PROBLEM 5

ANSWER:

For (a), (b), (c) see solution.
(d) 8

USES OF THE PROBLEM:

This problem illustrates the power of pattern recognition in problem solving. Impossible to solve on the calculator, it is easily solved by constructing tables and finding the pattern.

SOLUTION:

(a) n	3^n	(b) n	7^n	(c) n	4^n
1	3	1	7	1	4
2	9	2	49	2	16
3	27	3	343	3	64
4	81	4	2401	4	256
5	243	5	16807	5	1024
6	729	6	117649	6	4096
7	2187	7	823543	7	16384
8	6561	8	5764801	8	65536
9	19683	9	40353607	9	262144
10	59049	10	282475249	10	1048576

(d)   Examining the table for 3^n, we find the units digits repeating the pattern 3,9,7,1. That is, they have a cycle of four. To see where it is in the cycle, we divide by four. We find those exponents that give a remainder of one will have a units digit of three. Those giving a remainder of two will have a units digit of nine. Those giving a remainder of three will have a units digit of seven, while those giving a remainder of zero will have a units digit of one.

Examining the units digit for 7^n, we find a similar case. The repeating is 7,9,3,1 for the units digits. Upon division of the exponents by four, we observe that:

    a remainder of one always produces a unit digit of 7
    a remainder of two produces a units digit of 9
    a remainder of three produces a units digit of 3
    a remainder of zero produces a units digit of 1

Examining the units digits for 4^n, we find the repeating pattern is 4,6 for the units digits. Upon division of the exponents by two, we observe that a remainder of one always produces a units digit of four, while a remainder of zero always produces a units digit of six.

Thus, any whole number whose units digit is three will produce a pattern identical to the units digit of 3^n and the pattern will be similar for 7^n and 4^n. For 13^{841}, we find $841 \div 4 = 210$, remainder 1, and the units digit will be one. For 24^{617}, we find $617 \div 2 = 308$, remainder 1, and the units digit will be four. Adding the three units digits, $3 + 1 + 4 = 8$, the correct sum.

TEACHING SUGGESTIONS:

This problem may be extended by asking similar questions about other numbers. For example, what is the units digit in 2^{81}? Another possible extension is to check the tens digit to determine if a cycle exists.

PROBLEM 6

ANSWER:

6159 cm^3

USES OF THE PROBLEM:

This problem gives good practice in using a sketch to visualize a problem. It is good exercise in using the Pythagorean Theorem and square roots.

SOLUTION:

Consider the sketch.

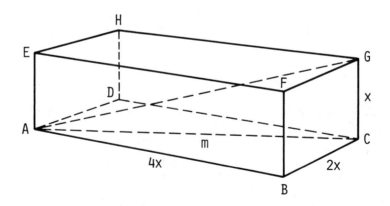

Space diagonal $\overline{AG}$ has a length of 42 cm and is the hypotenuse of right triangle ACG (let $\overline{AC}$ = m and let x = height).

Then $m^2 + x^2 = 42^2$ and $m^2 = 42^2 - x^2$

Also, $m^2 = (2x)^2 + (4x)^2 = 4x^2 + 16x^2 = 20x^2$

By the transitive property,

$$42^2 - x^2 = 20x^2$$

$$42^2 = 21x^2$$

$$\frac{42^2}{21} \text{ and } x^2 = \frac{42 \cdot 42}{21} = 2 \cdot 42 = 84$$

$$\text{and } x = \sqrt{84} = 2\sqrt{21}$$

The volume is length times width times height, so

$$2\sqrt{21} \cdot 4\sqrt{21} \cdot 8\sqrt{21} = 64 \cdot 21\sqrt{21} = 6159 \text{ cm}^3$$

TEACHING SUGGESTIONS:

Discuss the term "space diagonal." Ask students to point out one in the room (upper corner to opposite lower corner). Define the term "parallelepiped" and ask students for other names (e.g., rectangular prism).

PROBLEM 7

ANSWER:

35 ways

USES OF THE PROBLEM:

This problem encourages use of tree diagrams.

SOLUTION:

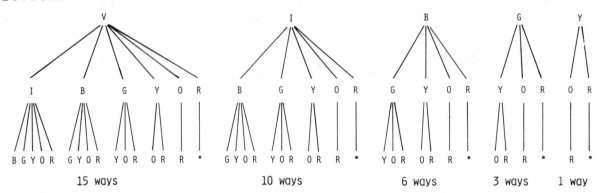

TEACHING SUGGESTIONS:

Encourage students to list possibilities. Visualizing the problem as a tree diagram will increase their ability to solve complex problems.

PROBLEM 8

ANSWER:

24 transformations

USES OF THE PROBLEM:

One purpose of problem solving is the development of strategies. In this problem, some students will write out each transformation until they return to the first one, but the teacher should encourage discovery of a strategy that will simplify the process.

SOLUTION:

Rather than making all the transformations, find out what happens to an individual position. (Let the arrows below represent one transformation.) Starting with 1 in position 1 (A1), we find:

1 goes to position 2	(1 → 2)
1 then goes to position 6	(2 → 6)
1 then goes to position 5	(6 → 5)
1 then goes to position 3	(5 → 3)
1 then goes to position 7	(3 → 7)
1 then returns to position 1	(7 → 1)

Thus, after six transformations, the 1 will return to its original position. Note that this will also be true for 2,6,5,3,7.

Starting with some number not in this series, such as 4, we find:

4 goes to position 15	(4 → 15)
4 then goes to position 14	(15 → 14)
4 then goes to position 9	(14 → 9)
4 then goes to position 13	(9 → 13)
4 then goes to position 12	(13 → 12)
4 then goes to position 16	(12 → 16)
4 then goes to position 11	(16 → 11)
4 then goes to position 4	(11 → 4)

Thus, after eight transformations, the 4 will return to its original position. Note that this will also be true for 15,14,9,13,12,16,11. Each of these numbers will return to its original position after eight transformations.

The 8 and 10 are stationary; they remain in their initial positions throughout the transformations.

There are two sequences, one of period six and the other of period eight. They will both return to their original positions after 24 transformations, the least common multiple.

TEACHING SUGGESTIONS:

This may be a confusing problem for some students. Using objects (such as pennies, checkers, poker chips) labelled with their starting numbers on grids like the ones below will help them visualize the process.

Go to 2	Go to 6	Go to 7	Go to 15
Go to 3	Go to 5	Go to 1	Stay
Go to 13	Stay	Go to 4	Go to 16
Go to 12	Go to 9	Go to 14	Go to 11

1	2	3	4
5	6	7	8
9	10	11	12
13	14	15	16

Objects numbered
1- 16

PROBLEM 9

ANSWER:

The four integers are 6,10,15, and 30.

USES OF THE PROBLEM:

This problem is useful for examining the properties of prime numbers, common factors, and prime factors.

SOLUTION:

The sum is to be a prime number and, therefore, an odd number, since it must be greater than two. This eliminates the set beginning with two. All other elements would require a multiple of two, and, thus, an even sum.

Similar reasoning eliminates three, for all elements would have a multiple of three, and the sum would be divisible by three. Four follows the reasoning for two. Five would be eliminated by the same process. Now, we try six. Six has as its factors two and three. Seven is not considered since it has no factors greater than one in common with six. Eight is 2·2·2 and follows the reasoning for either two or four. Nine is eliminated by the same reasoning used for three. Ten has factors of two and five, meeting the common factor requirement. The third number would require two of the three factors two, three, or five. Thus, 15 = 3·5. The fourth factor requiring two, three, or five would be 30 = 2·3·5. All numbers meet the requirement.

Checking, we get 6 + 10 + 15 + 30 = 61, which is prime. We have a solution.

TEACHING SUGGESTIONS:

This problem lends itself to a class discussion of prime numbers and strategies that would eliminate possibilities.

The problem could be modified to ask students to find other sets for which the given conditions would be true (e.g., 10,15,18,30).

PROBLEM 10

ANSWER:

(a) See table in solution.

(b) See table in solution.

(c) $S = \dfrac{n(n-1)}{2}$

USES OF THE PROBLEM:

This is another example of how illustrating a problem and tabulating results can reveal a pattern that solves the problem.

SOLUTION:

(a)   1 point   •
      0 segments
                      2 points   ●———●
                      1 segment
                                         3 points
                                         3 segments

4 points
6 segments
                      5 points
                      10 segments
                                         6 points
                                         15 segments

(b)

number of points	1	2	3	4	5	6	7	8	9	10
number of segments	0	1	3	6	10	15	21	28	36	45

(c)   Students may recognize the triangular numbers in part (b), or they may think of Gauss's theorem. The formula $\frac{n(n+1)}{2}$ does not produce the number of segments, but if we modify it to $\frac{n(n-1)}{2}$, it then fits the pattern.

TEACHING SUGGESTIONS:

Modify the problem. Instead of non-collinear points and line segments, it could be people at a party and the number of handshakes exchanged.

Have students draw a solution for 16, 24, or 32 points. The resulting designs are beautiful if the points are evenly spaced around a circle.

PROBLEM 11

ANSWER:

10,200

USES OF THE PROBLEM:

This problem is a good one for teaching the counting principle and
the use of a tree to record possibilities.

SOLUTION:

Consider five cards in sequence, such as ace, king, queen, jack, ten.
Each card in the sequence can be chosen in four ways. The total number of ways of forming such a sequence is 4^5 = 1024.

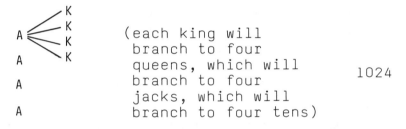

There are ten such sequences: ace (high), king (high), queen (high),
jack (high), ten (high), nine (high), eight (high), seven (high),
six (high), five (high). Note that a straight cannot begin with a
four as the high card.

Therefore, 1024·10 = 10,240 straights.

However, there are ten straight flushes for each suit, or 4·10 = 40
in all. Thus, 10,240 - 40 = 10,200 straights.

TEACHING SUGGESTIONS:

Do a few simple counting problems to illustrate the use of the tree.
For example, you could ask students to find the number of possibilities for seating three people in three chairs.

PROBLEM 12

ANSWER:

37 and 333,667

USES OF THE PROBLEM:

This is a good problem to use as a basis for exploration of prime
numbers and prime factorization.

SOLUTION:

Examining the product 111,111,111, we note that it is divisible by 111. The quotient is 1,001,001.

$$1,001,001 \div 3 = 333,667$$
$$111 \div 3 = 37$$

The product $3 \cdot 3$ is the prime factors of the multiplier, nine. Therefore, 37 must be a factor of 12,345,679.

We have $3 \cdot 3 \cdot 37 \cdot 333,667 = 111,111,111$.

Using a calculator, we can determine in a few minutes that the factor 333,667 is also prime. (To do this, try dividing 333,667 by the prime numbers up to the square root of 333,667.)

TEACHING SUGGESTIONS:

You might introduce this problem with a review lesson on divisibility. What is a quick check for divisibility by 2,3,5,9?

If you are testing a number to see if it is prime, look first to see that it is odd (two is the only even prime number), then do trial divisions for 3,5,7,11,13, etc. until you reach the greatest prime that is less than the square root of the number. For example, if testing 541, we can see it is not divisible by three (since the sum of its digits is not divisible by three), so we would proceed by testing successive primes up to 23, the greatest prime less than the square root. We find that 541 is prime.

PROBLEM 13

ANSWER:

11:04 AM

USES OF THE PROBLEM:

This problem is good for developing logical reasoning.

SOLUTION:

143,999,999,993 minutes is only seven minutes short of being 144,000,000,000 minutes, a number that is a multiple of 60. Therefore, $144,000,000,000 \div 60 = 2,400,000,000$ hours = 100,000,000 days, in which case the time would be 11:11 AM. Since the number given is seven short of that, the time will be 11:04 AM.

TEACHING SUGGESTIONS:

The numbers involved are too large for most calculators, so students will have to discover a way of simplifying the problem.

PROBLEM 14

ANSWER:

See solution, below.

USES OF THE PROBLEM:

This problem develops the ability to visualize solids and gives practice in organizing data, observing patterns, and deriving a formula for the general case. Also, it gives students the opportunity to use knowledge from other problems (problems six and ten) to assist them in solving the problem.

SOLUTION:

(a)   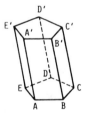   Lower base: ABCDE
                               Upper base: A'B'C'D'E'
                               Lateral edge: DD', CC', EE', AA', BB'
                               Lateral face: B'BCC', C'CDD', D'DEE',
                                              E'EAA', A'ABB'
                               The upper and lower bases are parallel.

(b)   Constructing a table,

number of lateral faces (n)	3	4	5	6	7	8	9	10	30
number of space diagonals (s)	0	4	10	18	28	40	54	70	810

      As the number of lateral faces increases by one, the number of space diagonals increases by four, then six, then eight, ten, twelve, etc.

(c)   The number of space diagonals is the number of all segments that can be formed, less all edges (lateral and base) and surface diagonals.

      If n is the number of lateral faces, then 2n is the number of vertices that can be connected by segments. Thus, in a five-

57

sided prism, there are ten vertices that can be connected by segments. Using the formula derived in problem 10, we see that the number of segments is $\frac{p(p-1)}{2}$ where p is the number of vertices. Since p = 2n, we get $\frac{2n(2n-1)}{2} = 2n^2 - n$ segments. The upper base has $\frac{n(n-1)}{2}$ segments (including diagonals), as does the lower base. Thus, there are $\frac{2n(n-1)}{2} = n^2 - n$ edges and diagonals in the top and bottom of the prism. Subtracting from the total, we get $2n^2 - n - (n^2-n)$, or n^2. From this we subtract the number of lateral edges, which are equal in number to the lateral faces, n. Hence, we have $n^2 - n$. Each lateral face is a parallelogram, and thus contains two diagonals, so we subtract 2n and we get $n^2 - n - 2n = n^2 - 3n$, the number of space diagonals.

TEACHING SUGGESTIONS:

Building models is helpful in visualizing the problem. A cardboard box can be used for the four-sided prism (four lateral faces) and string can be used for the space diagonals. Paper can be folded to form the lateral faces of the five-sided prism and pentagons cut to form the bases.

PROBLEM 15

ANSWER:

(a) 120   (b) 501

USES OF THE PROBLEM:

This problem is another that is good for teaching the use of a tree to count possibilities, though here an organized list is, perhaps, more useful in part (b).

SOLUTION:

(a) If no digits are repeated, there are five choices for the first digit, four choices for the second, three for the third, two for the fourth, and only one for the fifth. Thus, $5 \cdot 4 \cdot 3 \cdot 2 \cdot 1 = 120$.

(b) If only the five may be repeated, there are 501 possibilities,
     computed as follows.

                 five-digit numbers with one five: 120
                 five-digit numbers with two fives: 240
                 five-digit numbers with three fives: 120
                 five-digit numbers with four fives: 20
                 five-digit numbers with five fives: 1
                                                             ___
                                                             501

     Looking at the case of a five-digit number with two fives, we
     observe the following possibilities.

```
5 5 __ __ __      So, there are ten ways to arrange two fives.
5 __ 5 __ __
5 __ __ 5 __      The numbers 1,2,3,4 can be arranged in the
5 __ __ __ 5      three blanks in 24 ways.
__ 5 5 __ __
__ 5 __ 5 __      1 2 3    1 2 4    1 3 4    2 3 4
__ 5 __ __ 5      1 3 2    1 4 2    1 4 3    2 4 3
__ __ 5 5 __      2 3 1    2 1 4    3 1 4    3 2 4
__ __ 5 __ 5      2 1 3    2 4 1    3 4 1    3 4 2
__ __ __ 5 5      3 1 2    4 1 2    4 1 3    4 2 3
                  3 2 1    4 2 1    4 3 1    4 3 2
```

     Thus, there are 10·24 = 240 five-digit numbers with two fives.

     Three fives can be arranged in ten ways. The numbers 1,2,3,4
     can fill the remaining spaces in twelve ways. Thus, 10·12 =
     120 five-digit numbers with three fives.

     Four fives can be arranged in five ways. The numbers 1,2,3,4
     can be arranged in the remaining position in four ways. Thus,
     5·4 = 20 five-digit numbers with four fives.

     Five fives can be arranged in only one way.

TEACHING SUGGESTIONS:

Have your students make organized lists to chart the possibilities
of arranging the digits in the five spaces.

To demonstrate the idea of arranging four digits in two positions,
have students count the ways they could seat two students in two
chairs if they are choosing from a group of four. Then have them
arrange three in three chairs, choosing from a group of four.

PROBLEM 16

ANSWER:

10,100 tons

USES OF THE PROBLEM:

This measurement problem requires organization and care in computation.

SOLUTION:

We can think of the bathysphere as having 1000 cubic feet of water pushing down on it. So, on every square foot of surface, there will be a force of 64·3·1000, or 64,300 lb.

Since the surface area of a sphere is given by $4 \pi r^2$, the area of the bathysphere is:
$$4 \cdot \pi \cdot 5^2 = 100 \text{ square feet.}$$
The total force, then, is:
$$100 \pi \cdot 64,300 \text{ lb} = 20200440.76 \text{ lb, or } 10,100 \text{ tons.}$$

TEACHING SUGGESTIONS:

If your students don't know the formula for the surface area of a sphere, have them locate it; don't give it to them. You could include with the problem an assignment to do some research on bathyspheres. What is the greatest depth to which one has descended? What was the pressure per square foot at that depth?

PROBLEM 17

ANSWER:

200 gallons/2000 miles

USES OF THE PROBLEM:

This problem involves estimation and elimination of possibilities by trial and error.

SOLUTION:

The Cartesian truck had a range of 500 miles. Thus, the problem was how to get 50 gallons of gasoline stored 500 miles from the end point (300 miles from the starting point). A round trip to that point was not possible, for the total range would be 600 miles. Using reasonable estimation and experimentation, which are often useful in problem solving, the intrepid mathematician accomplished the trip in the following manner.

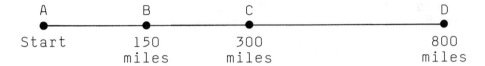

A	B	C	D
Start	150 miles	300 miles	800 miles

trips	gas used/miles
A→B; leave 20 gallons; B→A	30 gal., 300 mi.
A→B; leave 20 gallons; B→A	30 gal., 300 mi.
A→B; pick up 15 gallons; B→C; leave 20 gal.	30 gal., 300 mi.
C→B; pick up 15 gallons; B→A	30 gal., 300 mi.
A→B; pick up 10 gallons; B→C; pick up 20 gal.	80 gal., 800 mi.
TOTALS	200 gal., 2000 mi.

TEACHING SUGGESTIONS:

This is a good problem to use in the midst of some that have more precise solutions. Most students will have to roll up their sleeves and get their hands dirty to solve this one; that is, they'll have to make several trials to arrive at a correct solution.

PROBLEM 18

ANSWER:

(a) hexagon = 120

     heptagon = $128\frac{4}{7}$

     octagon = 135

(b) See solution.

(c) 170

(d) 40 sides

USES OF THE PROBLEM:

This is a good example of how a simple theorem in geometry (the sum of the angles in a triangle is 180) can be used to solve more complex problems.

SOLUTION:

(a) We can develop the proper patterns by considering the number of non-overlapping triangles within each figure. For example, we find in a pentagon that there are three non-overlapping triangles.

Each triangle contains 180 or $3 \cdot 180 = 540°$, the sum of the measures of the angles of a regular pentagon. Then, $540 \div 5$ angles = 108.

In similar fashion, a regular hexagon = 120°, a regular heptagon = $128\frac{4}{7}°$, and a regular octagon = 135°.

(b)

sides (n)	3	4	5	6	7	8	9	10	11	12	n
sum of all angles	180	360	540	720	900	1080	1260	1440	1620	1800	180(n-2)
measure of each angle	60	90	108	120	$128\frac{4}{7}$	135	140	144	147	150	$\frac{180(n-2)}{n}$

(c) $\frac{180(36-2)}{36} = 5(34) = 170$

(d) $171 = \frac{180(n-2)}{n}$

$171n = 180n - 360$
$360 = 9n$
$n = 40$

TEACHING SUGGESTIONS:

Review the concept of the sum of the angles in a triangle being equal to 180 (a graphic illustration is to cut out a triangle, cut off its vertices, then lay the pieces side by side. The result will always be a straight line).

You can extend the problem by having your students construct the polygons in parts c and d.

PROBLEM 19

ANSWER:

$\frac{11}{16}$

USES OF THE PROBLEM:

This problem is useful in distinguishing between a "counting tree" and a "probability tree." The problem can, however, be solved by either method.

SOLUTION:

For team A to win the championship they must win two more games.

Ways for Team A to win the championship

Game	1	2	3	4	5	Probability
	W	W	W	–	–	$\frac{1}{2} \cdot \frac{1}{2} = \frac{1}{4}$
	W	W	L	W	–	$\frac{1}{2} \cdot \frac{1}{2} \cdot \frac{1}{2} = \frac{1}{8}$
	W	L	W	W	–	$\frac{1}{2} \cdot \frac{1}{2} \cdot \frac{1}{2} = \frac{1}{8}$
	W	L	L	W	W	$\frac{1}{2} \cdot \frac{1}{2} \cdot \frac{1}{2} \cdot \frac{1}{2} = \frac{1}{16}$
	W	W	L	L	W	$\frac{1}{2} \cdot \frac{1}{2} \cdot \frac{1}{2} \cdot \frac{1}{2} = \frac{1}{16}$
	W	L	W	L	W	$\frac{1}{2} \cdot \frac{1}{2} \cdot \frac{1}{2} \cdot \frac{1}{2} = \frac{1}{16}$
					TOTAL	$\frac{11}{16}$

Thus, there are six ways for team A to win, assuming that A won the first game.

Since the teams are evenly matched, the probability tree would look like this:

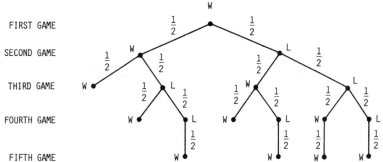

The paths through the tree that give team A the championship have probabilities of $\frac{1}{4} + \frac{1}{8} + \frac{1}{8} + \frac{1}{16} + \frac{1}{16} + \frac{1}{16}$ for a total of $\frac{11}{16}$.

TEACHING SUGGESTIONS:

Many students will decide the probability is $\frac{6}{10}$ since there are six ways for team A to win and ten possible outcomes for the five-game series, given that A won the first game. A look at the tree will show, however, that team A has a one-in-four chance of winning the championship in three games, a two-in-eight chance of winning it in four games, and a four-in-sixteen chance of winning it in five games (to see this clearly, extend the tree to five games even if there is a winner).

PROBLEM 20

ANSWER:

(a) N = 81

(b) N = 162

USES OF THE PROBLEM:

This problem is useful in exploring the patterns of prime factors and exponents.

SOLUTION:

(a)

sums of primes = 12	product of primes
2 + 2 + 2 + 2 + 2 + 2	$2^6 = 64$
2 + 2 + 2 + 3 + 3	$2^3 \cdot 3^2 = 72$
2 + 2 + 3 + 5	$2^2 \cdot 3 \cdot 5 = 60$
2 + 5 + 5	$2 \cdot 5^2 = 50$
3 + 3 + 3 + 3	$3^4 = 81*$
2 + 3 + 7	$2 \cdot 3 \cdot 7 = 42$
5 + 7	$5 \cdot 7 = 35$

sums of primes = 14	product of primes
2 + 2 + 2 + 2 + 2 + 2 + 2	$2^7 = 128$
2 + 2 + 2 + 2 + 3 + 3	$2^4 \cdot 3^2 = 144$
2 + 3 + 3 + 3 + 3	$2 \cdot 3^4 = 162*$
2 + 2 + 2 + 3 + 5	$2^3 \cdot 3 \cdot 5 = 120$
2 + 2 + 3 + 7	$3 \cdot 2^2 \cdot 7 = 84$
3 + 3 + 3 + 5	$3^3 \cdot 5 = 135$
2 + 5 + 7	$2 \cdot 5 \cdot 7 = 70$
7 + 7	$7 \cdot 7 = 49$

* largest number

TEACHING SUGGESTIONS:

This is another example of using a simple table to simplify what seems, on the surface, a complex problem.

Make sure your students are familiar with the concept of a function. A simple definition is that a function is a pairing, or matching up, of two sets so that each element in the first set corresponds with exactly one element in the second set.

You might have students experiment with other numbers and challenge them to make a conjecture as to the combination of primes that will produce the greater product.

PROBLEM 21

ANSWER:

(a)

n	1	2	3	4	5	6	7	8	9	10	11	12	13	14	15	16	17	18	19	20
F(n)	1	1	2	3	5	8	13	21	34	55	89	144	233	377	610	987	1597	2584	4181	6765

(b)

n	5	4	3	2	1	0	-1	-2	-3	-4	-5	-6	-7	-8	-9	-10	-11	-12	-13	-14	-15	-16
F(n)	5	3	2	1	1	0	1	-1	2	-3	5	-8	13	-21	34	-55	89	-144	233	-377	610	-987

(c) $F_{-n} = (-1)^{n-1} \cdot F_n$

USES OF THE PROBLEM:

This is a good problem to use as a lead-in to a study of other sequences (a sequence is an arrangement of terms such that each successive term follows the preceding one according to a uniform rule).

SOLUTION:

Complete the tables according to the rules given. Note that, among the negative values for n, F(n) is positive when n is odd and negative when n is even.

TEACHING SUGGESTIONS:

The Fibonacci sequence is one that appears in a wide variety of things--from plants to music. For an interesting discussion of its history and applications, see Chapter Two of Mathematics, A Human Endeavor, by Harold Jacobs.

Explore other sequences (arithmetic, geometric, square, cubic, etc.) and have students state the rule for each.

PROBLEM 22

ANSWER:

(a) 360°

(b) See solution, below

USES OF THE PROBLEM:

This problem should be used after Problem 18 (measure of angles in polygons). It involves use of simple geometric concepts (supplementary angles, definition of isosceles triangles) and prior experience.

SOLUTION:

We shall consider all n-gons, excluding the triangle and square.

In the case of the regular hexagon, we can first look back to Problem 18. The measure of each angle was found to be 120°. The triangle formed by the intersection of the sides extended is always isosceles or equilateral, since the base angles are equal. In this case, the base angles are 60°, this being the supplement of 120°. The angle at the point of the star, then, is 180° - 2x.

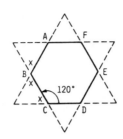

There are six such angles, so we have the sum of the angles as 6(180 - 2x) = 6(60) = 360°. Using the same process we can complete the table.

n vertices	5	6	7	8	9	10	n
measure of each star angle	36	60	$77\frac{1}{7}$	90	100	108	$\frac{180(n-4)}{n}$
sum of the star angles	180	360	540	720	900	1080	180(n-4)

TEACHING SUGGESTIONS:

We suggest you not refer to Problem 18; it should occur to your students. You could review the properties of isosceles triangles and supplementary angles.

PROBLEM 23

ANSWER:

(a) 216   (b) 10,11   (c) $\frac{25}{216}$ (d) $\frac{5}{9}$ (e) $\frac{20}{27}$

USES OF THE PROBLEM:

This problem reveals patterns in the counting of outcomes and requires a systematic organization of data.

SOLUTION:

(a) There are 216 possible outcomes when three dice are rolled. These can be counted by drawing a "tree," making an organized list, or by use of fundamental counting principles (i.e., since there are six choices for the first die, six for the second die, and six for the third, there are 6·6·6, or 216 outcomes).

(b) There are 27 ways to roll either a ten or an eleven. For example,

<u>Sum of ten</u>

136	235	145	226	244	334
163	253	154	262	424	343
316	325	415	622	422	433
361	352	451			
613	523	514			
631	532	541			

Similarly, there are 27 ways to roll a sum of eleven.

(c) There are 25 ways to roll a nine, so the probability is $\frac{25}{216}$.

(d) There are 120 ways to roll three different numbers: six choices for the first die, five for the second, and four for the third.

So, there are 6·5·4 = 120 ways, and $\frac{120}{216} = \frac{5}{9}$.

(e) There are 160 ways to get a sum less than twelve, so the probability is $\frac{160}{216} = \frac{20}{27}$. The outcomes for each sum are:

sum of 3 dice	possible outcomes	sum of 3 dice	possible outcomes
3	1	11	27
4	3	12	25
5	6	13	21
6	10	14	15
7	15	15	10
8	21	16	6
9	25	17	3
10	27	18	1

TEACHING SUGGESTIONS:

Have your students organize their data and look for patterns. Many students will predict that the possible outcomes for nine are 28, since, up to that point, the possible outcomes seem to be the triangular numbers.

Remind students that the sum of the possible outcomes must match their answer to part a.

PROBLEM 24

ANSWER:

215, 216, 217

USES OF THE PROBLEM:

This is a good illustration of the concept of least common multiple.

SOLUTION:

Trying successive multiples of five, we find that the first two quantities satisfying the requirement for five and six are 35 and 36.
Further experimentation indicates that more such quantities are found
by adding 30. This gives such pairs as 65,66 and 95,96. If we consider six and seven, we find 48 and 49 as the first pair satisfying
the condition. Adding 42, we find additional pairs, such as 90,91
and 132,133.

Thus, by examining a simpler problem containing only two of the
numbers, we find that, by determining common multiples, we need only
add the two numbers and we find quantities that satisfy the condition. For example, for five and six, the least common multiple is
30. Adding five and six, successively, we get 35,36.

The least common multiple for five, six, and seven is 210. By adding
the five, six, and seven successively to 210 we determine the required answer to be 215, 216, and 217.

TEACHING SUGGESTIONS:

Introduce the problem by discussing the concept of least common multiple, but don't use the term at this point (save that for discussion
of solutions). For example, you might ask students to list pairs of
numbers that are divisible by three and four, respectively, up to
100.

Make sure your students understand what is meant by "respectively,"
i.e., that we divide the first number by three, the second by four.

PROBLEM 25

ANSWER:

(a) even   (b) odd   (c) even   (d) even   (e) odd

USES OF THE PROBLEM:

This problem is useful in increasing understanding of place value.

SOLUTION:

Since the place values in Base 7 are odd--1,7,49,343, etc.--if there
is an odd digit in the place its value will be odd, and if there is
an even digit in the place its value will be even. Furthermore,
since an odd plus an odd makes an even, and odd times even is even,
if the sum of the digits is even, the number is even. If the sum
is not even, the number is odd.

If the number is divisible by eight, the sum of the digits in the
even places (i.e., 2nd, 4th, etc.) minus the sum of the digits in
the odd places is divisible by eight. For example, consider the
Base 7 number 462. There is a six in the second place, and the sum
of the digits in the first and third places is six. Thus, 6 - 6 = 0,
which is divisible by 8.

This is a very difficult pattern to observe. You may wish to give
students a hint concerning the odd and even digits. Examine the
table of multiples of eight in Base 7.

11	110	206	305	404
22	121	220	316	415
33	132	231	330	426
44	143	242	341	440
55	154	253	352	451
66	165	264	363	462

You will notice that the difference between the sum of the even
digits and odd digits is either zero or eight. Try larger numbers
to determine if the pattern holds. For example,

$$16063_7 \div 11_7 = 1433_7$$

$$\text{and } (6+6) - (1+0+3) = 12 - 4 = 8$$

70

TEACHING SUGGESTIONS:

Introduce this problem by discussing what makes a number odd or even and what happens when you add and multiply odd and even numbers in all possible ways.

PROBLEM 26

ANSWER:

(a)

row	1	2	3	4	5	6	n
no. of terms	3	5	7	9	11	13	2n + 1

(b) See solution, below.

USES OF THE PROBLEM:

This problem is useful in demonstrating the power of pattern recognition in problem solving. It also gives practice in multiplication of polynomials.

SOLUTION:

ROW 1	1	1	1	Sum is 3
ROW 2	1	2	3	2	1	Sum is 9
ROW 3	1	3	6	7	6	3	1	Sum is 27
ROW 4	1	4	10	16	19	16	10	4	1	Sum is 81
ROW 5	1	5	15	30	45	51	45	30	15	5	1	Sum is 243
ROW 6	1	6	21	50	90	126	141	126	90	50	21	6	1	Sum is 729
ROW 7	1	7	28	77	161	266	357	393	357	266	161	77	28	7	1	Sum is 2187

The sums are all powers of 3, from 3^1 to 3^7. It would be expected that the sum of the elements in the nth row would be 3^n.

The coefficients in the algebraic expansions correspond to the numbers in the row that is the same as the exponent. Thus, $(x^2+x+1)^6 =$

$$x^{12}+6x^{11}+21x^{10}+50x^9+90x^8+126x^7+141x^6+126x^5+90x^4+50x^3+21x^2+6x+1$$

TEACHING SUGGESTIONS:

If you have not previously introduced the Pascal Triangle, this would be a good time. There are many comparisons that can be made between this pattern and Pascal's triangle. For example, the sum of the numbers in each row of the Pascal triangle is a power of two. In this table the sum is a power of three.

PROBLEM 27

ANSWER:

404 feet

USES OF THE PROBLEM:

This problem reinforces the idea that making a sketch can be helpful
in problem solving. It also offers practice with the Pythagorean
Theorem and simple equations.

SOLUTION:

Consider the diagram below. Let A represent the position of the ship
at the time it drops anchor and B its position four hours later. Let
x be the length of the anchor chain.

In right triangle BCO, $x^2 = 80^2 + (x-8)^2$

$$\text{and} \quad x^2 = 6400 + x^2 - 16x + 64$$
$$16x = 6464$$
$$x = 404 \text{ ft}$$

So, 404 ft is the amount of chain in the water.

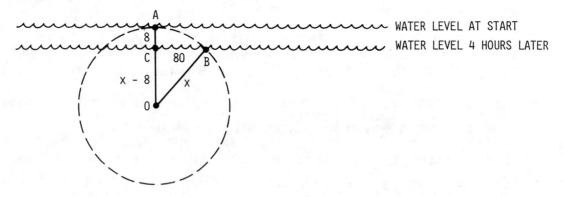

TEACHING SUGGESTIONS:

If you use this problem with students who have not had algebra, we
suggest that you teach them to square x - 8 by the multiplication

rules they know, without resort to rules of algebra.

That is,
$$
\begin{array}{r}
x - 8 \\
\underline{x - 8} \\
- 8x + 64 \\
\underline{x^2 - 8x} \\
x^2 - 16x + 64
\end{array}
$$
rather than $(a-b)^2 = a^2 - 2ab + b^2$

PROBLEM 28

ANSWER:

$\frac{5}{6}$

USES OF THE PROBLEM:

This problem defies the intuitive guess of most students $(\frac{1}{2})$ and demonstrates the usefulness of a tree.

SOLUTION:

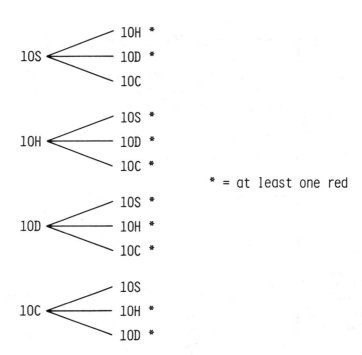

* = at least one red

Thus, the probability is $\frac{10}{12} = \frac{5}{6}$

73

TEACHING SUGGESTIONS:

You might introduce the problem by having students make a guess
about the probability. Most will guess that it's one-half.

Have them draw a tree or make an organized list. Then it will be
apparent that far more than half the possible outcomes contain a
red ten.

PROBLEM 29

ANSWER:

(-50, -50)

USES OF THE PROBLEM:

This problem is good practice in following directions, plotting
coordinates, and observing patterns.

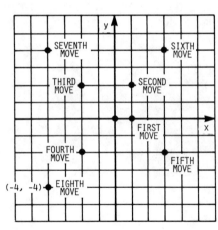

SOLUTION:

After four moves the position is (-2,-2).
We note that in a sequence of four moves,
x decreases by two and y decreases by two.
After eight moves the position is (-4,-4).
x has decreased by four and y has decreased
by four. After twelve moves the position is
(-6,-6). So, after 100 moves the position
must be (-50,-50).

TEACHING SUGGESTIONS:

We suggest you have your students plot the first twelve points. Some
students will, of course, see the pattern by then. Those who don't
should keep plotting.

You could extend the problem by having students make up rules for
other progressions of points. (One further extension is found in
problem 37.)

PROBLEM 30

ANSWER:

45 cm^2

USES OF THE PROBLEM:

This problem demonstrates the value of making a sketch. Cutting out
the rectangles (making a model) would be an even more graphic solution.

SOLUTION:

Note that if the rectangles are attached only at a corner, we see

the difference is 80 cm^2 - 35 cm^2 = 45 cm^2. (See figure 1.) Now

slide the smaller rectangle such that there is an overlap of 1 cm^2.

The non-overlapping difference is 79 cm^2 - 34 cm^2 = 45 cm^2. (See

figure 2.)

Continuing this process will reduce each area by an equal value

and the difference will remain 45 cm^2. The final step shows the

smaller rectangle as entirely enclosed by the larger, and the diff-

erence of the non-overlapping regions is 45 cm^2 - 0 = 45 cm^2. (See

figure 3.)

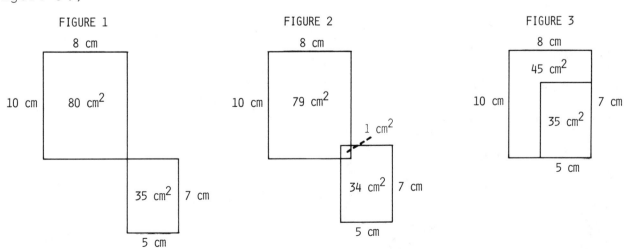

TEACHING SUGGESTIONS:

Use graph paper and have the students cut out the rectangles. Students could work in groups, sliding the rectangles around and recording the area of the non-overlapping regions (which, of course, will always be 45 cm^2).

PROBLEM 31

ANSWER:

See solution, below

USES OF THE PROBLEM:

Pattern recognition and organization of data are necessary to the solution of this problem.

SOLUTION:

(a)
```
1    2
1    3    2
1    4    5    2
1    5    9    7    2
1    6   14   16    9    2
1    7   20   30   25   11    2
1    8   27   50   55   36   13    2
1    9   35   77  105   91   49   15    2
1   10   44  112  182  196  140   64   17    2
1   11   54  156  294  378  336  204   81   19    2
```

(b) The upward diagonals (left to right) are:

1,3,4,7,11,18,29,47,76,123,...

(c) 199,322,521,843,1364

(d) The sums of the first seven rows are:

3,6,12,24,48,96,192,...

This suggests the following pattern:

$$3 \cdot 1 = 3 \cdot 2^0$$

$$3 \cdot 2 = 3 \cdot 2^1$$

$$3 \cdot 4 = 3 \cdot 2^2$$

$$3 \cdot 8 = 3 \cdot 2^3$$

$$3 \cdot 16 = 3 \cdot 2^4$$

$$3 \cdot 32 = 3 \cdot 2^5$$

$$3 \cdot 64 = 3 \cdot 2^6$$

This pattern leads us to the formula for finding the sum of the terms in any given row, n: $3 \cdot 2^{(n-1)}$

TEACHING SUGGESTIONS:

If your students are not familiar with the Pascal triangle, you can either show them how it is constructed or let them discover its pattern. We suggest the latter. The problem can, of course, be solved without knowledge of the Pascal triangle.

This is a sequence often associated with the Fibonacci sequence. It is called the Lucas sequence and is formed in the same manner, by adding the last two terms of the sequence to get the succeeding term.

PROBLEM 32

ANSWER:

See solution, below

USES OF THE PROBLEM:

This problem affords excellent practice in using the calculator as an aid in problem solving. It also gives good practice in the concept of square root.

SOLUTION:

(a) The minimum value that can be assigned to x is $\frac{1}{2}$. Any lesser value creates the square root of a negative real number, which is not possible under the conditions given.

if $x = \frac{1}{2}$, then $\sqrt{\frac{1}{2} + \sqrt{(1-1)}} + \sqrt{\frac{1}{2} - \sqrt{(1-1)}} =$

$\sqrt{\frac{1}{2} + 0} + \sqrt{\frac{1}{2} - 0} = \sqrt{\frac{1}{2}} + \sqrt{\frac{1}{2}} = \frac{\sqrt{2}}{2} + \frac{\sqrt{2}}{2} = \frac{2\sqrt{2}}{2} = \sqrt{2}$

Note that the smallest value of A is $\sqrt{2}$.

(b) $A = \sqrt{2}$ for all real values where $x \geq \frac{1}{2}$ and $x \leq 1$.   (This is an

excellent opportunity for students to reason to an answer through use of a calculator.)

(c) $\frac{3}{2}$ or 1.5

Young students will probably use the "guess-and-check" method of finding a solution for this part.

Recalling part a, we know the value of x must be greater than one. The first guess that we might try is x = 2. We find that if A is to have a value of 2, then x = 2 gives 2 < 2.4494896, and x must be between 1 and 2. We could next try 1.5.

We find, $2 = \sqrt{1.5 + \sqrt{3 - 1}} + \sqrt{1.5 - \sqrt{3 - 1}}$
$ 2 = 1.7071067 + .29289332$

Try it on your calculator.

If you wish to see an algebraic solution for this problem, we refer you to Problem 2 of the International Mathematics Olympiad, 1959. (Greitzer, Samuel L. International Mathematical Olympiads 1959-1977. Washington, D.C.: The Mathematical Association of America, New Mathematical Library, 1978.)

TEACHING SUGGESTIONS:

An algebraic solution is too difficult for most students in grades 7-9, but in playing around with the problem and a calculator, we found the delightful solution printed here, which the youngsters were able to do and then "rationalize" why it worked.

PROBLEM 33

ANSWER:

$$\frac{10259}{2187}$$

USES OF THE PROBLEM:

This problem offers good practice in one of the techniques of problem solving--thinking of a simpler example.

SOLUTION:

This problem must be carefully examined. It is most easily understood by assuming the tank of fresh water holds one gallon. Any other amount will be proportional.

Let x represent the pounds of salt per gallon in the tank at any given time. Then, it follows that each time we pour off $\frac{1}{3}$ of the tank we will add $\frac{5}{3}$ pounds of salt in filling the tank. (Thinking in terms of one gallon, adding $\frac{1}{3}$ gallon of the brine solution that has five pounds of salt in it will add $\frac{5}{3}$ pounds of salt.)

Step 1: (a) We pour off $\frac{1}{3}$ gallon of the water.

(b) Remaining is $\frac{2}{3}$ multiplied by x, the number of pounds of salt in the tank at the time. Thus, we have $\frac{2}{3} \cdot (0) + \frac{5}{3} = \frac{5}{3}$ pounds of salt per gallon.

Step 2: (a) We pour off $\frac{1}{3}$ of the mixture, and remaining is $\frac{2}{3} \cdot \frac{5}{3} = \frac{10}{9}$ pounds of salt.

(b) On filling, we add $\frac{5}{3}$ pounds to $\frac{10}{9}$, or $\frac{5}{3} + \frac{2}{3} \cdot \frac{5}{3} = \frac{5}{3} + \frac{10}{9} = \frac{25}{9}$ pounds of salt per gallon.

Step 3: (a) We pour off $\frac{1}{3}$, and remaining is $\frac{25}{9} \cdot \frac{2}{3}$, or $\frac{50}{27}$ pounds of salt per gallon.

(b) On filling, we add $\frac{5}{3}$ pounds to $\frac{50}{27}$ to give us $\frac{95}{27}$ pounds of salt per gallon.

Step 4: $\frac{5}{3} + \frac{95}{27}\cdot\frac{2}{3} = \frac{325}{81}$ pounds of salt per gallon.

Step 5: $\frac{5}{3} + \frac{325}{81}\cdot\frac{2}{3} = \frac{1055}{243}$ pounds of salt per gallon.

Step 6: $\frac{5}{3} + \frac{1055}{243}\cdot\frac{2}{3} = \frac{3325}{729}$ pounds of salt per gallon.

Step 7: $\frac{5}{3} + \frac{3325}{729}\cdot\frac{2}{3} = \frac{10259}{2187}$ pounds of salt per gallon.

TEACHING SUGGESTIONS:

We suggest you lead students through the first step. The problem will not be difficult if they carefully keep track of the steps.

PROBLEM 34

ANSWER:

2.33 in.

USES OF THE PROBLEM:

This problem provides practice in computing the volume of a cylinder.

SOLUTION:

The volume of a cylinder is given by $\pi r^2 h$. The new can will be five inches high. If x is its radius, its volume is $\pi x^2 \cdot 5$. The volume of the old can was $\pi \cdot 1.3^2 \cdot 4$. The volumes are equal, so we get:

$$\pi x^2 \cdot 5 = \pi \cdot 1.3^2 \cdot 4$$
$$5x^2 = 6.76$$

$$x^2 = 1.352$$
$$x = 1.1627553 \text{ inches}$$

The diameter, then, is 2.33 in., rounded to hundredths.

TEACHING SUGGESTIONS:

As in many problems involving the use of pi, it is not necessary to use values for pi. Here, they divide out.

The problem could be extended by changing the dimensions.

PROBLEM 35

ANSWER:

$$3\sqrt{3} + 3 - \frac{3\pi}{2}$$

USES OF THE PROBLEM:

This is a good review of the area of circles and is good practice in simple equations.

SOLUTION:

If the area of each circle is $\frac{\pi}{4}$, then the radius of the circle is determined as:

$$\frac{\pi}{4} = \pi r^2 \Rightarrow r^2 = \frac{1}{4} \Rightarrow r = \frac{1}{2}$$

Thus, the length $\overline{BC}$ is $\frac{1}{2} + \frac{1}{2} + \sqrt{3} = 1 + \sqrt{3}$

The diameter of each circle is 1 and the length $\overline{AB}$ is 3.
The area of rectangle ABCD is $3(1 + \sqrt{3}) = 3 + 3\sqrt{3}$.

The area of each circle is $\frac{\pi}{4}$. The area of the six circles is $\frac{\pi}{4} \cdot 6 = \frac{3\pi}{2}$.

The area of rectangle ABCD not covered by the circles is
$3 + 3\sqrt{3} - \frac{3\pi}{2}$.

TEACHING SUGGESTIONS:

Junior high students are sometimes motivated by knowing they are solving problems that are intended for older students. This problem is based on one that was on the S.A.T.

PROBLEM 36

ANSWER:

(a) 9   (b) 9   (c) 315054 442323 221535   (d) 6   (e) 6

USES OF THE PROBLEM:

This problem is useful for examining divisibility rules and number bases other than ten.

SOLUTION:

Choose any integer (for example, 98761). When the digits are reversed and you subtract it from the first, you always get a multiple of nine (in the example below, 8 + 1 + 9 + 7 + 2 = 27).

$$
\begin{array}{r}
98761 \\
-16789 \\
\hline
81972
\end{array}
$$

In Base 7, reversing the digits and subtracting gives a number that is divisible by six and whose digits give a sum that is divisible by six. For example,

$$
\begin{array}{r}
436021_7 \\
-120634_7 \\
\hline
315054_7
\end{array}
$$

Dividing in Base 7, we get

$$
\begin{array}{r}
35343 \\
6\,\overline{)315054} \\
24 \\
\hline
45 \\
42 \\
\hline
30 \\
24 \\
\hline
35 \\
33 \\
\hline
24 \\
24 \\
\hline
\end{array}
$$

TEACHING SUGGESTIONS:

Encourage your students to do the Base 7 computations without converting to Base 10. This would be a good time to review rules of divisibility.

PROBLEM 37

ANSWER:

(0, -100)

USES OF THE PROBLEM:

Following directions, plotting coordinates, and observing patterns
are the problem-solving skills used in this problem.

SOLUTION:

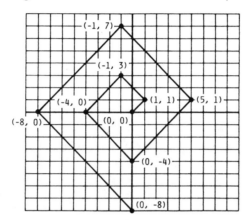

Step no.	Coordinates
1	(1,1)
2	(-1,3)
3	(-4,0)
4	(0,-4)
5	(5,1)
6	(-1,7)
7	(-8,0)
8	(0,-8)

For the fourth step one arrives at
(0,-4). For the next four steps the
x changes by 5-6-7+8=0, while y
changes by 5+6-7-8=-4. Similarly,
for each four steps the x value will
be 0, while the y value will decrease
by 4.

Thus, the final point is (0, -100).

TEACHING SUGGESTIONS:

As in Problem 29, students should plot at least the first twelve
points.

Extend the problem by having students create new conditions for
plotting successive points.

PROBLEM 38

ANSWER:

See solution, below.

USES OF THE PROBLEM:

This problem offers excellent practice in place value and
numeration.

SOLUTION:

In Base 10 the pattern is an eight-digit number times the highest
digit in that base, nine. In Base 4, therefore, we would try 13·3,
which gives us a product of 111. Look at this multiplication: 13·3.
We multiply three by three and get nine. Since we are in Base 4, we
take out as many fours as we can (two) and count them in the fours
place. This leaves one in the ones place. We then multiply the
three by the one in the fours place and add the two fours we carried
over. This gives five. We take out as many fours as we can (one)
and put them in the four squared, or sixteens, place. This leaves us
with a one in the fours place. Thus, our product is 111.

Base	Relation
4	13·3 = 111
5	124·4 = 1111
6	1235·5 = 11111
7	12346·6 = 111111
8	123457·7 = 1111111
9	1234568·8 = 11111111
10	12345679·9 = 111111111
11	12345678A·A = 1111111111
12	123456789B·B = 11111111111

Base	Relation
4	13·12 = 222
5	124·13 = 2222
6	1235·14 = 22222
7	12346·15 = 222222
8	123457·16 = 2222222
9	1234568·17 = 22222222
10	12345679·18 = 222222222
11	12345678A·19 = 2222222222
12	123456789B·1A = 22222222222

84

$$\begin{array}{r} 13_4 \\ \times 12_4 \\ \hline 32_4 \\ 13 \\ \hline 222_4 \end{array} \qquad \begin{array}{r} 124_5 \\ \times 13_5 \\ \hline 432_5 \\ 124 \\ \hline 2222_5 \end{array} \qquad \begin{array}{r} 1235_6 \\ \times 14_6 \\ \hline 5432_6 \\ 1235 \\ \hline 22222_6 \end{array} \qquad \begin{array}{r} 12346_7 \\ \times 15_7 \\ \hline 65432_7 \\ 12346 \\ \hline 222222_7 \end{array}$$

TEACHING SUGGESTIONS:

We suggest that you have your students do addition and multiplication in other bases before they do this problem. Ideally, they should stay within the base and not convert the problem to Base 10.

PROBLEM 39

ANSWER:

144

USES OF THE PROBLEM:

This problem is useful for teaching the relationship between the area of a circle and its radius.

SOLUTION:

The cross-sectional area of the one-foot diameter pipe is πr^2, where the radius is six inches. The area, then, is 36π. The cross-sectional area of the one-inch diameter pipe is $\pi(.5)^2$, or $.25\pi$. Dividing 36π by $.25$ gives 144, so the larger pipe can carry 144 times as much water as the smaller. Thus, it would take 144 of the one-inch pipes to equal the large pipe.

TEACHING SUGGESTIONS:

As in many problems involving the use of pi, it is helpful not to use a numerical approximation of pi until the last step, if necessary. Here, it is not necessary because pi divided by pi is one.

PROBLEM 40

ANSWER:

5.20 inches and 7.94 inches

USES OF THE PROBLEM:

This is good practice in the area of circles and in relating the parts to the whole. It provides the opportunity to consider a simpler problem.

SOLUTION:

r_1 = 3 inches   Let A_1 = the area of the inner circle

r_2 = radius of second circle A_2 = area of the 50-point region

r_3 = radius of outer circle A_3 = area of the 25-point region

$A_1 = 9\pi$ in.^{2} Since A_2 is worth half as much as A_1, its area, being inversely proportional, should be twice that of A_1, or 18π in.^{2}

Thus, $A_2 = \pi r_2^2 - 9\pi$ in.^{2} and, $\pi r_2^2 - 9\pi$ in.$^2 = 18\pi$ in.^{2}

$$\pi r_2^2 = 27\pi \text{ in.}^2$$

$$r_2 = 5.20 \text{ in.}$$

$$A_3 = \pi r_3^2 - \pi r_2^2 = \pi r_3^2 - 27\pi \text{ in.}^2$$

Since A_3 is worth half as much as A_2, its area should be twice that of A_2, or 36π in.^{2}.

$$\pi r_3^2 - 27\pi \text{ in.}^2 = 36\pi \text{ in.}^2$$

$$\pi r_3^2 = 63\pi \text{ in.}^2$$

$$r_3 = 7.94 \text{ in.}$$

TEACHING SUGGESTIONS:

Make sure your students understand the idea of inverse proportion.

Most will see the idea of subtracting the area of the inner circles from that of the outer to find the area of the bands, but you could introduce this problem with a simpler one. For example, what is the area of a border around a rectangular picture?

PROBLEM 41

ANSWER:

$9500 at 11% and $2500 at 8.5%

USES OF THE PROBLEM:

This problem may be used to practice algebraic concepts. Students who have not had the necessary algebra may solve it by the guess-and-check method.

SOLUTION:

First, by algebra: Let x = amount at 11%

$$.11x + .085 (12000 - x) = \$1257.50$$
$$.11x + 1020.00 - .085x = 1257.50$$
$$.025x = 237.50$$
$$x = \$9500.00$$

So, the amount at 8.5% is $12000 - 9500 = $2500

A guess-and-check solution might look like this:

   Guess: $6000 at 11%, $6000 at 8.5%
           .11(6000) + .085(6000) = 660 + 510 = 1170
Since this amount is too small, try increasing the amount at 11%.
   Guess: $8000 at 11%, $4000 at 8.5%
           .11(8000) + .085(4000) = 880 + 340 = 1220
Since this guess is still too small, the next guess might be $10000 and $2000.

By evaluating the result of each guess, the student can narrow down the possibilities until the correct answer is reached.

TEACHING SUGGESTIONS:

Students who use the guess-and-check method should definitely use calculators.

The problem may be easily extended by varying the principal, the rates, and the time.